The Shaggiest Dog Story Ever Told

-or-

The Story of Dark Matter

by

Rustin K. Miller

LIONHEART GROUP PUBLISHING
Printed in the USA

The Shaggiest Dog Story Ever Told or The Story of Dark Matter

by Rustin K. Miller

Paperback ISBN: 978-1-938505-52-2
E-Book ISBN: 978-1-938505-53-9

10 9 8 7 6 5 4 3 2 1
First Edition ~ April 2020

Published by Lionheart Group Publishing, Colorado, USA
Printed in the USA

Visit us on the web at www. lionheartgrouppublishing.com

Dedicated to my loving sister Sandi, whose perseverance is an inspiration to all who know her. Many thanks for her invaluable help in publishing my own booklet.

Intro

I'LL START MY STORY WITH an inarguable cosmic constant. This is: 59-year-old *neo-duffers* without formal scientific schoolin' ought not try to explain *Dark Matter*, or invent cosmologies.

However, if history, science, and several Giants in a few disciplines may [someday] forgive me, I might provide a chuckle or two, if not a use-able concept or notion.

I certainly do not believe that I have a viable *Theory of Everything* (T.O.E.), or that I have all the questions answered. I only offer an hypothesis for what the Dark Matter is, or could be. At this point it seems reasonable, to me, that informed imagination divided by Occam's Razor = My guess could be as good as anybody's.

Not apart from the neo-duffer bit, I like to fancy myself an "*Amateur Conceptual Cosmology En-*

thusiast." This doesn't really mean much but it is a satisfyingly long title, and may provide some cover for the assumptions I am about to make.

For the most part, I won't disturb the standard model at all, squeezing my story between the Big Bang and condensation (or precipitation) of the Quark/gluon soup that filled the early universe.

The Brief Story of Dark Matter

"Imagination is more important than knowledge. Knowledge is limited. Imagination encircles the world."

Albert Einstein

THIS IS WHAT I'VE IMAGINED for the Dark Matter:

First, all at once, there was nothing! From our perspective, anyway. Then, the Big Bang happened, and…

My first assumption is that our universe 'erupted' from an unstable Kaluza-Klein Universe (KKU). I visualize this KKU rotating 'open' in several dimensions and burping out a vast quantity of energy in an effort to stabilize itself and find a happy compactification state. This bubble *might* have broken away to inflate and allowed the KKU to continue its rotation toward a new compactified state, but that didn't happen.

Instead, as this new bubble began to expand, it filled with the *Monopole Field* and the *Polaron Field*. The Monopole Field is comprised of monopoles and the Polaron Field is comprised of polarons (my term).

Monopoles, for my purpose, are point-like, empty vessels which like to capture polarons. When three or more polarons are captured in a monopole, the magnetic moment of the monopole is determined by the sum of the polarons (example: two north polarons and one south polaron = north monopole). Monopoles may capture any number of polarons, and the magnetic moment is north or south, depending on whichever is present in greater numbers. Monopoles are happiest with just three polarons, but the abundance of polarons far exceeds the population of monopoles in the very early universe. Therefore, most monopoles in this period are 'pumped up' with dozens, or even hundreds (thousands?) of polarons at any given atto-second. I realize that this sounds like the monopoles are non-existent entities and that the polarons would more-or-less be the monopoles, but this is the difficulty with relating my notion.

Another way to view this is that the monopole is a point in space/time (5D and/or 6D) where polarons may orbit without scattering. This point in space/time is perhaps the intersection of two higher dimensions where polarons are temporarily captured and engage in their complex orbits to form the monopole.

Polarons are force carriers and come in two flavors. North, with a clockwise spin, and south, with an anti-clockwise spin. Polarons interact only with other polarons, monopoles, and gravity. The only two interactions possible between polarons are: 1) two polarons in the environment will try to flip one another's spin during scattering, I call it *Ruskin T. scattering*, and 2) polarons orbit one another in the monopole in complex orbits, which I fancy to call, *Karl Misler orbits*.

There are four conditions of the polaron:

1) environmental polarons are polarons just zooming thru space and not associated with a monopole,

2) captured polarons are polarons (temporarily) captured in a monopole,

3) confined polarons are polarons confined in a monopole. And,

4) *orphaned* polarons, or *abandoned* polarons are polarons excluded from monopoles after polaron confinement. These are the same as environmental polarons, but with nowhere to go (can't be captured anymore).

Confinement and abandonment do not happen until the end of the Monopole/Polaron Field era, and only during *The Folding* and *The Pairing*.

"I am enough of an artist to draw freely upon my imagination."

Albert Einstein

The Monopole Field and the Polaron Field are interactive and monopoles are capturing (absorbing) and emitting polarons at a furious rate, each monopole flipping from north to south thousands or millions of times per atto-second. There is no ontological state of north or south for any monopole at this time as energies are so high and flipping rates so rapid that each flipped state is a virtual state. Therefore, all monopoles are effectively |mostly south, mostly north| all the time.

This gives rise to a force which tries to collapse the monopole field into a giant mono-nucleus. This force is countered and equaled by the force of the KKU trying to rotate onward towards a happier compactification state, and an equilibrium is established, keeping the KKU open and the bubble caught in its throat. The space/time bubble couldn't expand and break free because the monopoles it held were tethered in the one or two higher dimensions which tried to open with the first four (3D plus time) dimensions, and wanted to recompactify with the KKU. Nor could the KKU sever the bubble and continue rotating towards compactified. I call this *"The Holding."*

The Holding, The Big Chance, The Folding, and The Pairing

THIS CONFIGURATION HELD STEADY IN a meta-stable state for a good, long time, maybe thirty seconds...

...maybe 42.6 trillion years in The Holding. It seemed that nothing would break the status-quo (no expansion, no cooling).

Then, by sheer chance alone, *The Big Chance*, the law of non-zero probabilities came into play (anything has a non-zero chance of happening, eventually). During The Holding, monopoles were distributed throughout the field randomly, but somewhat evenly.

When the non-zero probability arose wherein all the monopoles on the 'right side' of our bubble (so to speak) were north at the same instant, and all the monopoles on the 'left side' were south for that same instant, it locked into that configuration for long enough to have an ontological state of |all north one side, all south the other side|.

This generated a tremendous magnetic potential which folded the two halves of the Monopole Field together in what I have dubbed "The Folding." This was like a swooshing together of all the individual monopoles, one north with one south, in what I call "The Pairing."

I picture it as relatively slow and graceful, Grand and Stately, so I think of it as "*The Folding Waltz*," resulting in The Pairing. The Pairing is simply when the monopoles 'paired up'.

THE SQUARING DANCE, DARK PARTICLES, AND THE TURNING AWAY (ABANDONMENT)

"I believe in intuitions and inspirations.
I sometimes feel that I am right.
I do not know that I am."

Albert Einstein

T**HE SQUARING DANCE IS A** difficult concept to relate. When I picture this event, I suspect (intuit) that these monopoles could not pair up during a simple folding. In a simple folding the monopoles would all act as fermions and try to stick together as di-pole mono-pairs, and we would have little magnets for (not) dark matter. These are not permitted due to an exclusionary principle.

As I watch my own movie in my head, I see one half of the Monopole Field rotating by 180 degrees with respect to the other half, or each half rotating 90 degrees with respect to one another and in opposite directions. This, I call "The Squaring Dance."

This Squaring Dance reorients the north half with the south half such that these monopoles now come together as Bosons, and can lay completely one over (inside) the other and their magnetic potentials become confined in our newly formed Dark Particles, which I fancy to call, *"Dark Magnets"*—my term for the Dark Matter.

At the moment of The Pairing, in what I call, *"The Turning Away,"* each monopole abandoned, or orphaned all but three of the polarons that it had captured the moment before, such that each Dark Magnet now contained a total of six polarons—three north and three south—which orbit inside the paired monopoles in complex *Karl Misler orbits*.

The higher dimensions could now release their tethers on the monopoles. The orphaned polarons were emitted into the environment and the space/time bubble we are so fond of finally broke free and drifted away from the now re-compactified KKU.

Rustin K. Miller's **Law of Conservation of Di-Pole Magnets**

Because of The Folding and The Pairing of the Monopole Field, the symmetry of the Monopole Field was broken and a new symmetry was established. We live in a universe of di-pole magnets because we sorta live between the two halves of the new symmetry!

This gave rise to the ***Law of Conservation of Di-Pole Magnets****.*

"The important thing is to not stop questioning. Curiosity has its own reason for existing."

Albert Einstein

Polaron Sublimation, The Higg's Field, and Expansion

With this new symmetry established the orphaned polarons had nowhere to go and their environment had changed (minus higher dimensions). This left them with nothing to do except to sublimate (or precipitate, or condense, freeze out, boil, whatever…melt?) into the quark gluon (baryonic) soup.

Polarons are the largest quanta of energy allowed, sort of like super-hyper-mega photons. The Polaron Field 'melted' into the matter/energy fields we find today.

When polarons scatter (*Ruskin T scattering*) in the pre-folded universe, they simply reverse one-another's spin and continue on their way.

This scattering isn't symmetrical, though. Many times only one polaron will flip its spin, or both will, and then one will flip back to its original spin, resulting in a large disparity between north and south polaron populations.

At the moment of The Turning Away, there happened to be many more north polarons than south polarons. During sublimation, the north polarons melted into matter, and the south polarons melted into antimatter. Then, the story unfolds as previously described by real cosmologists.

This is when *The Higg's Field* was created, giving rise to mass for the polarons. The Dark Magnets are so massive simply because there are six massive polarons within them. All polarons not confined by monopoles became baryonic matter/energy, and all monopoles were paired during The Pairing. The only monopole possible in our universe now would be the odd one out, ***if*** there were an uneven number of monopoles in the beginning.

The Higg's Field was left as a marker for the confined polarons, and as space/time continues to expand, The Higg's Field will slowly and *mysteriously* siphon off just enough energy to inform the new space/time with its minimum vacuum energy level requirement.

Therefore, The Dark Magnets are slowly getting weaker, and in some distant epoch (42.6 trillion years from now, perhaps) they might simply break apart and reform (by chance, occasionally) to make very weak di-pole (not-dark) magnets and The Higg's Field will be nearly zero. This, I suppose, could simply be called de-flation, or deflation of The Higg's Field.

"If I have seen further, it is that I have stood on the T.O.E.'s of Giants."

Rustin K. Miller

SUMMARY

IN MY EPIC, I BELIEVE that the monopoles did not inflate away. I like to think that they hung around, preventing the recompactification of the KKU and the expansion of the new space/time bubble, and simply held this new universe in a meta-stable state for an unknowable period of time.

This status-quo couldn't resolve by way of cooling, or expansion, and had to wait for The Big Chance wherein everything lined up |all north one side, all south the other side|. This is when The Folding happened and all the monopoles, save possibly one, paired up and formed Dark Magnets.

At this moment, the *Law of Conservation of Dipole Magnets* was established along with The Higg's Field, and sublimation of the Polaron Field. Then, our universe broke free of the KKU, and began its expansion. The Polaron Field melted to become our matter/anti-matter field and supplied the minimum zero-point energy requirement for the empty space which already existed.

The Higg's Field regulates the flow of energy from the polarons confined in The Dark Magnets to inform newly expanded space/time with its zero-point energy, and may, in some very distant age, dissipate entirely.

CONCLUSION

I DO NOT LAY CLAIM HERE to a theory, in that I cannot falsify these ideas. That should and will be left to those who know how to do that. I am only proposing what I feel is a possible scenario in the realm of imagination.

I don't think I'm the (Schrodinger's) cat's meow/no meow. I'm just a getting'–old–guy with a new idea. One that's maybe worthy of casual consideration...or at least a few deep belly laughs.

And who knows? The law of non-zero probabilities clearly holds that I could be spot on...

AFTERWARD

THIS BOOKLET IS JUST AN idea, or an hypothesis, for what the dark matter is, and is published as food for thought. It is a set of ideas which I feel are elegant (if not clumsily described) and perhaps even beautiful (my own bias, undoubtedly), and it only tries to explain:

What really happened with inflation,

What the Dark Matter is,

How Dark Particles formed in the very early universe,

Why Dark Particles are so massive,

What maybe happens to Dark Particles at the end of the universe,

Why The Higg's Field is there,

Explains one (1) Law of Conservation (which one....?),

Where baryonic matter comes from, and

Maybe explains why our observed vacuum energy is so low.

All this and a tiny bit more.

If this sounds ambitious, it didn't begin that way. I simply had an idea for what the dark matter could be and these other points kind of "fell out" on the way toward describing it. Some of them surprised me as much as they may the reader.

It's my hope that this will be read with an open and curious mind. I certainly don't mean to ruffle any feathers or offend any sensibilities. I'm not a scientist, and so this will not read as a scientific treatise. I apologize for any awkwardness in prose or form. I simply hope to convey the idea.

So, please read and enjoy my short story, and send me any criticism you may feel inspired to rage (or praise) on about. Thank you.

Rus

Glossary

A

Amateur Conceptual Cosmology Enthusiast: Me, a name I call myself. I claim this title as a neutral, non-scientific moniker to provide a modicum of cover for outrageous assumptions I make in the text of my short story on Dark Matter.

Abandonment (*Abandoned*): One of several terms used to describe how excess polarons are rejected by the monopoles during a transition period known as *The Folding* and *The Pairing*. See also: *Orphaned* and *The Turning Away*.

B

The Big Chance: The inevitable chance occurrence of all monopoles aligning in a state of: | all north on one side, all south on the other side |

as dictated by the law of non-zero probabilities (i.e. anything has a non-zero chance of happening, eventually).

C

Captured polarons: Polarons temporarily captured in a monopole. While captured, they orbit one another furiously and are emitted as quickly as new polarons are captured.

Confined polarons: Polarons which are captured in the instant before *The Folding*, and which are not orphaned with the excess polarons. Instead, they are permanently confined within the new *Dark Magnet*.

Confinement: The trapping of six polarons in a newly formed *Dark Magnet*. These are polarons which are kept by the *Dark Particle* and not abandoned to the environment.

D

Deflation: Deflation of The Higg's Field over vast time span due to the gradual siphoning-off of polaron energy confined in the *Dark Magnet*. This energy is distributed to the newly formed space/time created by the expansion of our universe.

Dark Magnets: When two monopoles pair up and abandon all but six polarons between them, their magnetic moment is forever confined within the dark particle along with their polarons, thus forming a *Dark Magnet*. *Dark Magnet* is my name for this dark particle.

Dark Particles: Same as *Dark Matter* or *Dark Magnets*. Two monopoles have paired up and confined their polarons to create a bit of dark matter.

E

Environmental polarons: Polarons just zooming through the void and not associated with a monopole. They interact with one another only through *Ruskin T. scattering*.

F

The Folding: The breaking of the Monopole Field symmetry resulting in *The Pairing* and *The Law of Conservation of Di-Pole Magnets*.

The Folding Waltz: The graceful and relatively slow coming together of all north monopoles with all south monopoles ('cept maybe one) resulting in *The Pairing*.

H

The Holding: The relatively long period when opposing forces prevented both expansion and contraction, leaving a status-quo wherein nothing could expand or cool, thereby leaving any change to wait for *The Big Chance*. During *The Holding,* the state of the monopole field remained in a meta-stable state of |mostly north, mostly south| .

K

Karl Misler orbits: The only other kind of interaction polarons can have with one another is in the monopole where they orbit furiously in complex, and as yet undefined ***Karl Misler orbits***.

L

Law of Conservation of Di-Pole Magnets: When the symmetry of the Monopole/Polaron Fields was broken during *The Folding*, the new symmetry established by *The Pairing* resulted in the new *Law of Conservation of Di-Poles magnets* as we sorta live between the two halves *cum* one again.

M

Monopole Field: An electro-magnetic field comprised of charged monopoles. Monopoles with an abundance of captured north polarons are north polarons and monopoles with a captured south abundance are south monopoles. Monopoles can capture any number of polarons but are happiest with only three. The monopoles, coupled with the polarons, form the monopole/ polaron field, or the *Monopole Field*.

N

Neo-duffer: Any getting'–old–guy, such as myself, who is new to getting' old. Another name I call myself!

O

Orphaned polarons: Any polaron abandoned to the environment during confinement, and left with nowhere to go (can't be captured any longer). See: *Abandonment*.

P

The Pairing: The joining of a north monopole with a south monopole during *The Folding*. All monopoles paired up during *The Folding* ('cept maybe one) at the end of *The Folding Waltz*.

Polaron Field: An electromagnetic field comprised of polarons. The polaron field just is the changing charge of the Monopole Field. Polarons are force carriers for the monopole and are not unlike the gluon for the quarks. They are nothing like the electrons. North polarons have a clockwise spin while south polarons possess an anti-clockwise spin. The polaron field only interacts with itself in *Ruskin T. scattering* in the environment, and by way of complex *Karl Misler orbits* within the monopole.

Polaron Sublimation: The newly *orphaned* polarons at the end of the pairing were left with nowhere to go and nothing to do. As the new space/time bubble expanded, this polaron flux sublimated into the ordinary matter we can detect around us. The polarons with a clockwise spin, (north polarons) became matter, while anti-clockwise (south) polarons became antimatter. There existed, at that moment, a huge disparity in north vs. south polarons, due to the asymmetrical *Ruskin T. scattering*, and we ended up with much more matter than antimatter.

R

Ruskin T. scattering: Polarons only interact

with one another in the environment by way of scattering. This scattering is simply when two polarons interact they flip one another's spin. If two like polarons come together, they flip to two south polarons and continue on their way. If two opposite polarons come together, the north flips to south and the south flips to north and they go their merry ways. This is not symmetrical in every case, and only one may flip its spin, or both will—but one will flip back to its former spin after the encounter. This leads to large disparities in populations of north vs. south polarons at various times.

S

The Squaring Dance: When *The Folding* and *The Pairing* occurred, the monopoles could not come together as fermions and thus required a rotation by each half of 90 degrees in opposite directions so that they could pair up as Bosons. This enabled them to 'nestle' together and confine both their polarons and their magnet moments and form *Dark Magnets*.

T

The Turning Away: During this event, monopoles expelled, or *abandoned* their excess polarons and kept only three per monopole, or six per dark magnet. This resulted in our *Dark Magnets* and the sublimation of the remaining polaron flux.

Made in the USA
Coppell, TX
06 November 2026

11491040R00023